MAPPING SETTLEMENTS

www.heinemann.co.uk/library
Visit our website to find out more information about Heinemann Library books.

To order:
Phone 44 (0) 1865 888066

Send a fax to 44 (0) 1865 314091

Visit the Heinemann Bookshop at www.heinemann.co.uk/library to browse our catalogue and order online.

First published in Great Britain by Heinemann Library, Halley Court, Jordan Hill, Oxford OX2 8EJ, part of Harcourt Education.
Heinemann is a registered trademark of Harcourt Education Ltd.

Editorial: Lucy Thunder and Harriet Milles
Design: Ron Kamen and Celia Jones
Illustrations: Barry Atkinson, Darren Lingard and Jeff Edwards
Picture Research: Melissa Allison and Beatrice Ray
Production: Camilla Smith

Originated by Repro Multi Warna
Printed and bound in China by WKT Company Limited

The paper used to print this book comes from sustainable resources.

10 dig ISBN 0 431 01324 1 (hardback)
13 dig ISBN 978 0 431 01324 4
09 08 07 06
10 9 8 7 6 5 4 3 2 1

10 dig ISBN 0 431 013292 (paperback)
13 dig ISBN 978 0 431 01329 9
09 08 07 06
10 9 8 7 6 5 4 3 2 1

British Library Cataloguing in Publication Data

Spilsbury, Louise
(Mapping the UK). – Mapping Settlements
526

A full catalogue record for this book is available from the British Library.

Acknowledgements
The Publishers would like to thank the following for permission to reproduce photographs:
Corbis Royalty-free pp. 24 (Buenos Aires), 25 (Sydney); GetMapping p. 8; Getty Images/Photodisc pp. 12 (motorway), 24 (New York), 25 (Paris, Tokyo); Harcourt Education Ltd/ Tudor Photography p. 26; Harcourt Education Ltd/Martin Sookias p. 12 (cemetery); Harcourt Education Ltd/Peter Evans p. 12 (railway station, city, bus station, church); Harcourt Education Ltd/Steve Benbow p. 12 (car park); Jane Hance pp. 7b, 12 (phone box); London Aerial Photo Library pp. 7t, 18; London Ambulance Service p. 12 (hospital); Philip's Maps p. 21; Photomap pp. 10, 11; Reproduced by permission of Ordnance Survey on behalf of The Controller of Her Majesty's Stationery Office, © Crown Copyright 100000230 pp.15, 17, 19, 20; Science Photo Library p. 23; Wimborne Minster Model Town pp. 4, 5.

Cover photograph of North Belfast reproduced with permission of Harcourt Education Ltd/Peter Evans. Section of Ordnance Survey map reproduced by permission of Ordnance Survey of Northern Ireland.

The Publishers would like to thank Dr Margaret Mackintosh, Honorary Editor of *Primary Geographer*, for her assistance in the preparation of this book.

Contents

Words appearing in the text in bold, **like this**, are explained in the Glossary.

▶ Look out for this symbol! When you see it next to a question, you will find the answer on page 29.

What are maps?

Maps are drawings of an area of the Earth, usually on flat pieces of paper. There are maps of small areas, such as a single park, or large areas like whole countries. Maps show us where places are, and the features we can find there. They can also help us find our way around a new area.

Hi, I'm Carta and I'm always getting lost in towns and cities! I'm coming with you on this great mapping adventure to find out how maps can help us.

Points of view

Have you ever visited a model village or town? There you can see models of roads, buildings and parks that look the same as they do in real life, only much smaller. When you walk round a model village, you look down on it from above. This is called an **aerial** view. Next time you are in a town, you could try looking down through a window from the top of a high building. You will get an aerial view of the streets and shops – and this is how maps view the world.

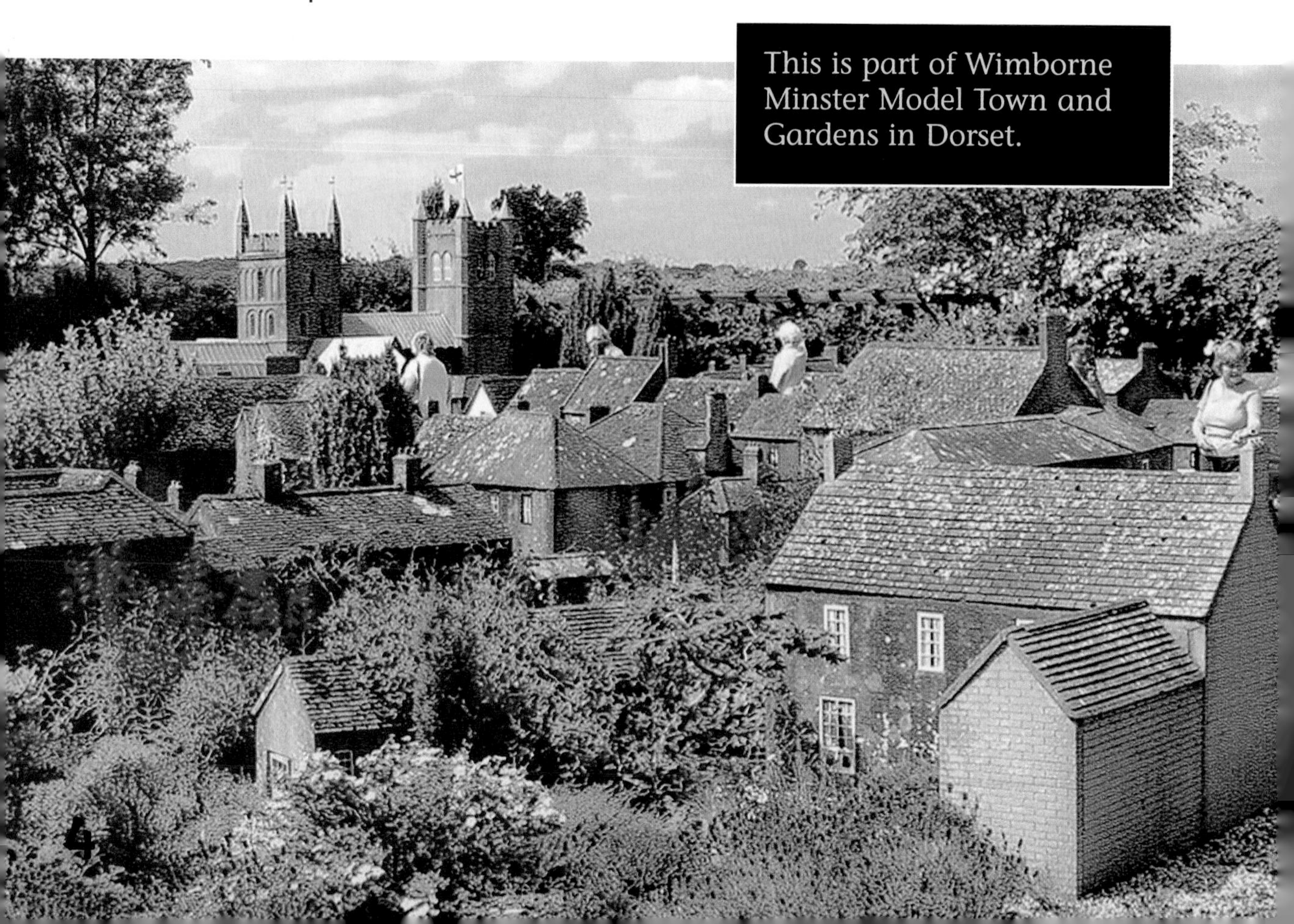

This is part of Wimborne Minster Model Town and Gardens in Dorset.

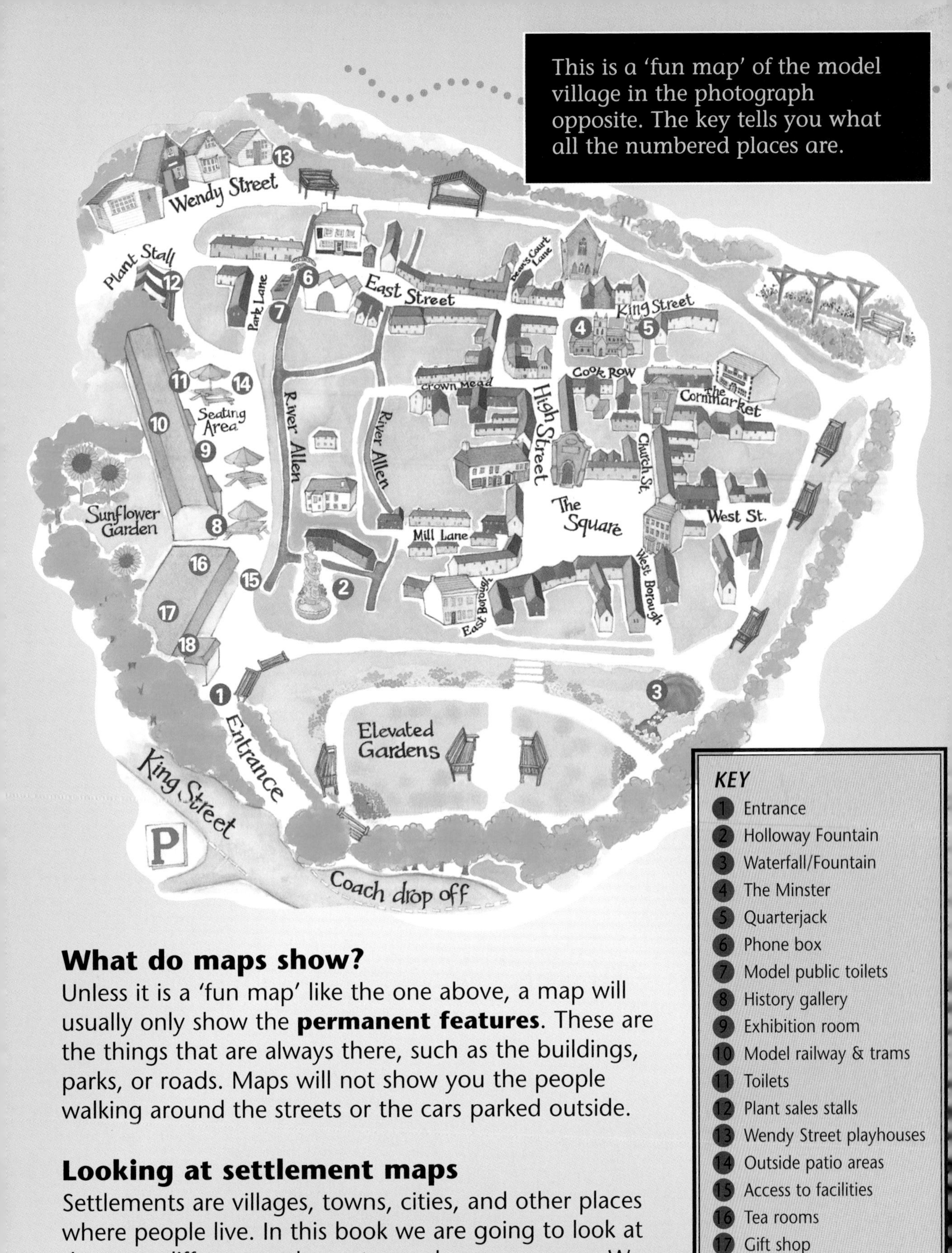

This is a 'fun map' of the model village in the photograph opposite. The key tells you what all the numbered places are.

KEY

1. Entrance
2. Holloway Fountain
3. Waterfall/Fountain
4. The Minster
5. Quarterjack
6. Phone box
7. Model public toilets
8. History gallery
9. Exhibition room
10. Model railway & trams
11. Toilets
12. Plant sales stalls
13. Wendy Street playhouses
14. Outside patio areas
15. Access to facilities
16. Tea rooms
17. Gift shop
18. Exit

What do maps show?

Unless it is a 'fun map' like the one above, a map will usually only show the **permanent features**. These are the things that are always there, such as the buildings, parks, or roads. Maps will not show you the people walking around the streets or the cars parked outside.

Looking at settlement maps

Settlements are villages, towns, cities, and other places where people live. In this book we are going to look at the ways different settlements are shown on maps. We will also use maps to find out about the important features of different settlements.

Looking at settlements

If you flew in a helicopter over the UK, you would see settlements of many different shapes and sizes. Yet all of them would have many features in common. This is because, wherever you go, people want and need similar things in order to live.

What makes a settlement?

Imagine you are using a computer game to build your own city. What would you put in it? Real settlements include buildings for people to live in, such as houses and flats. Then there are schools, offices, and factories where people go to learn or work. There are also shops, garages, hospitals, and libraries that provide **services** for people. Cities also have recreation areas, such as sports centres, museums, swimming pools, parks, and gardens. Finally, settlements need roads, railway lines, and cycle or pedestrian paths so that people can get around.

Different kinds of settlements

Settlements develop in a particular **site** for a reason. Seaside resorts are settlements on the coast near beaches. They usually have hotels and attractions for tourists who want to visit the sea. **Industrial** cities grew up around factories, and the areas of houses where the factory workers live. These settlements have a lot of railway and road links to transport goods and people. Market towns developed as meeting places for farmers to buy and sell goods or animals.

Settlements are always changing. They may get bigger, or even change their function. For example, Barnsley in Yorkshire was a coal-mining town. However, the last of the Barnsley coal mines closed in 1994. These days, the main jobs in the town are in building or factories.

This is an **aerial** photograph of the city of Plymouth in Devon. Plymouth is a **nucleated settlement.** It has lots of houses, shops, and other buildings grouped together around a centre, or nucleus.

This is a farming settlement in the Lake District in Cumbria. It is called a **dispersed settlement** because there are only a few buildings in a large area of countryside.

Marking settlements on a map

How do mapmakers make maps? They often start with an **aerial** photograph. People use cameras to take aerial photographs from aeroplanes or **satellites**. Aerial photographs can show whole settlements from above.

A seaside settlement

The aerial picture on this page shows Tenby in Wales. You can see that the town is built by a sheltered **harbour**. The harbour fills with water at high tide. Tenby developed as a port over 500 years ago. Ports are places where goods can be brought into a country or sent out to other countries by ship. In the past, Tenby was a busy port. Today, modern ships are very big and they stop at larger ports. Now Tenby's main **industry** is **tourism**.

Look closely at this aerial picture of Tenby and the map opposite. What clues can you see that tell you why tourists like to come here?

The name of a settlement often gives us clues about where it is! For example, Avonmouth is at the **mouth** of the river Avon.

From photo to map

The map on this page is a sketch map of Tenby made from the aerial photograph opposite. The map uses colours and shapes to show what the different features are. The blue area is the sea, and the sandy beaches are shown in yellow. The dark grey lines are the main roads and the buildings are shown in pink. Can you see the harbour wall in the map and the photograph?

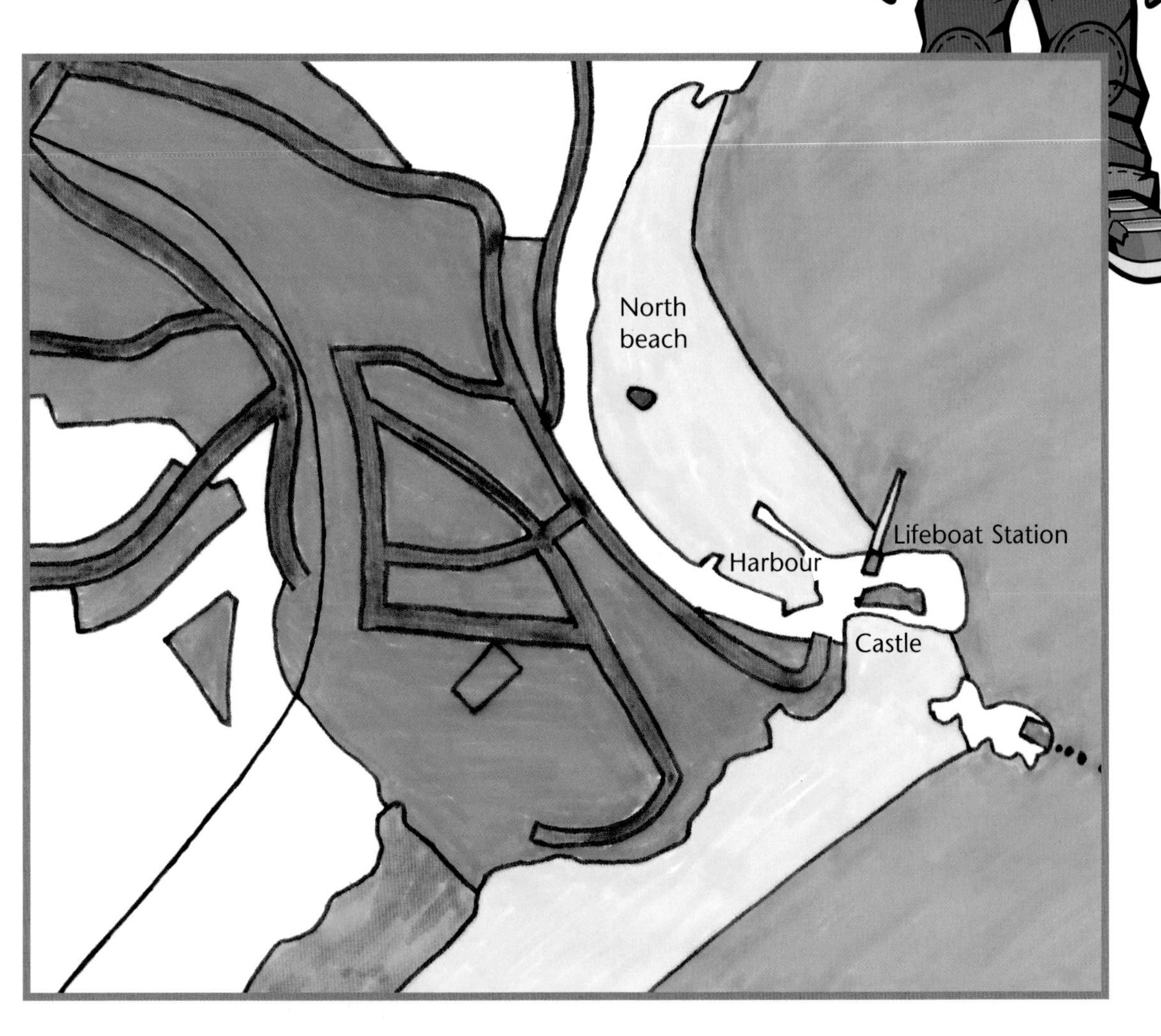

Exploring Cardiff

Cardiff is the capital city of Wales. It was once the biggest coal-exporting port in the world. At its busiest time, in 1913, more than 13 million tons of coal were carried in ships from Cardiff. Then the coal-mining **industry** of Wales was either slowed down or stopped altogether. The port of Cardiff was used less and less for exporting coal. The bay of Cardiff has now become a holiday resort, with lots of leisure activities on and off the water.

Aerial Cardiff

The **aerial** photograph on this page looks straight down on Cardiff from above. It is quite tricky to identify the things on the ground. From straight above, all the buildings look like rectangles or squares. Imagine you were visiting Cardiff for the first time, and you wanted to go to the sports stadium in the middle of the city. It would be difficult to find your way around just by using this picture!

This is a bird's-eye view of the centre of Cardiff.

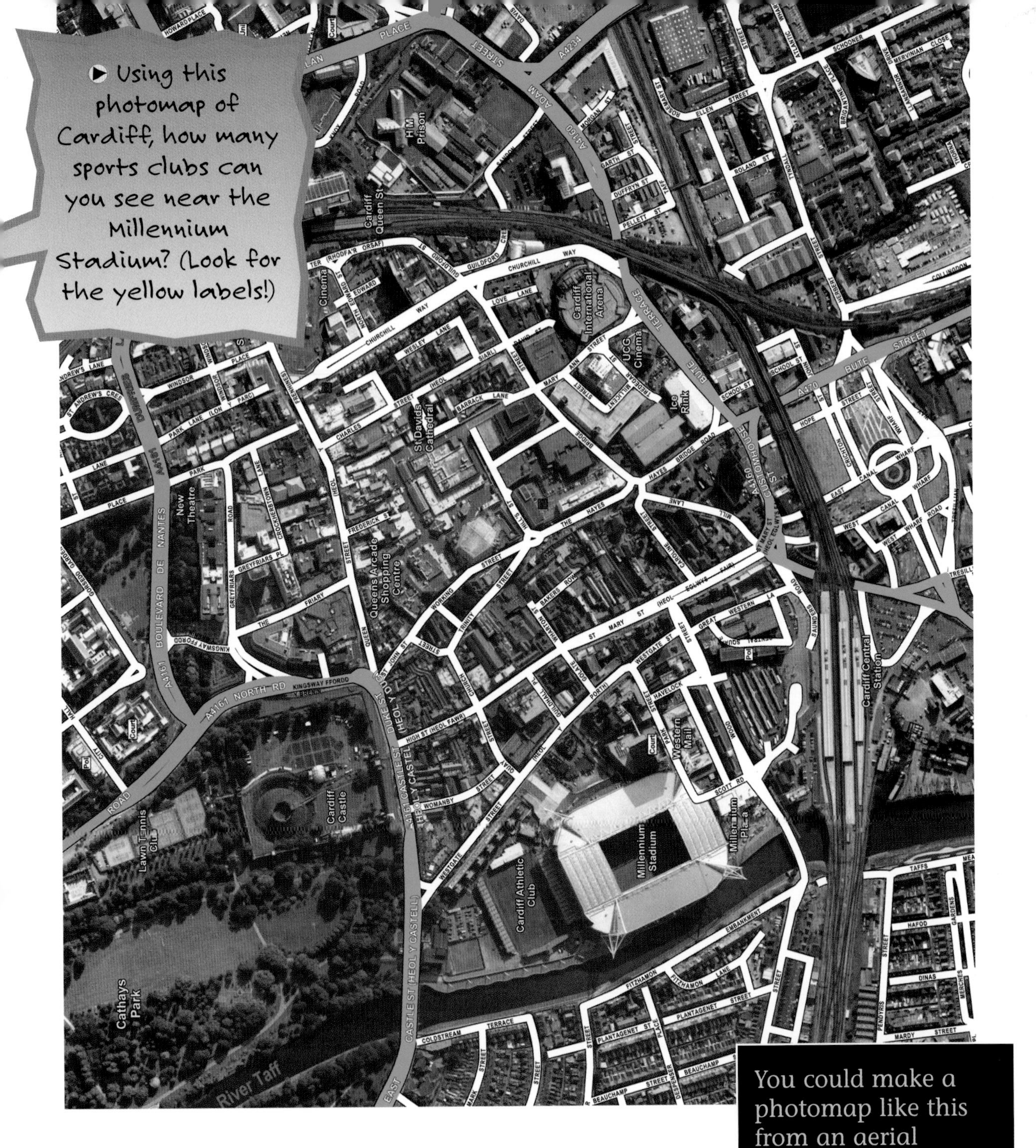

You could make a photomap like this from an aerial photograph of your home town. Try laying a sheet of tracing paper over the photo. Then draw in the names of any streets and buildings that you know.

A photomap

The map on this page is a special kind of map called a photomap. It uses the aerial photograph of Cardiff, and adds colours and labels for the streets and some of the places that visitors might want to find. The main roads through the city are coloured green and the smaller streets are white. These details make it easier for people to find their way around.

Settlement symbols

On some maps there are labels to tell us what the different features are – such as hospitals or stations. However, many maps use symbols for these features instead.

When a road goes under a railway line on a map, the line of the road is broken – but the line of the railway is continued. That shows you that the railway line goes over a bridge at that point. You can see an example of this on the map on page 13.

What are symbols?

Symbols are little pictures, coloured shapes, lines, patterns, or even letters that represent real things. We use symbols all the time. When a teacher gives you a star, it is a symbol to show that you have done good work! Black and white stripes on a Zebra crossing are symbols to show you where to cross safely. On a map, one symbol usually represents all the features of the same kind. For example, the map symbol for a car park is the same for all car parks in the UK, no matter how different the real car parks look.

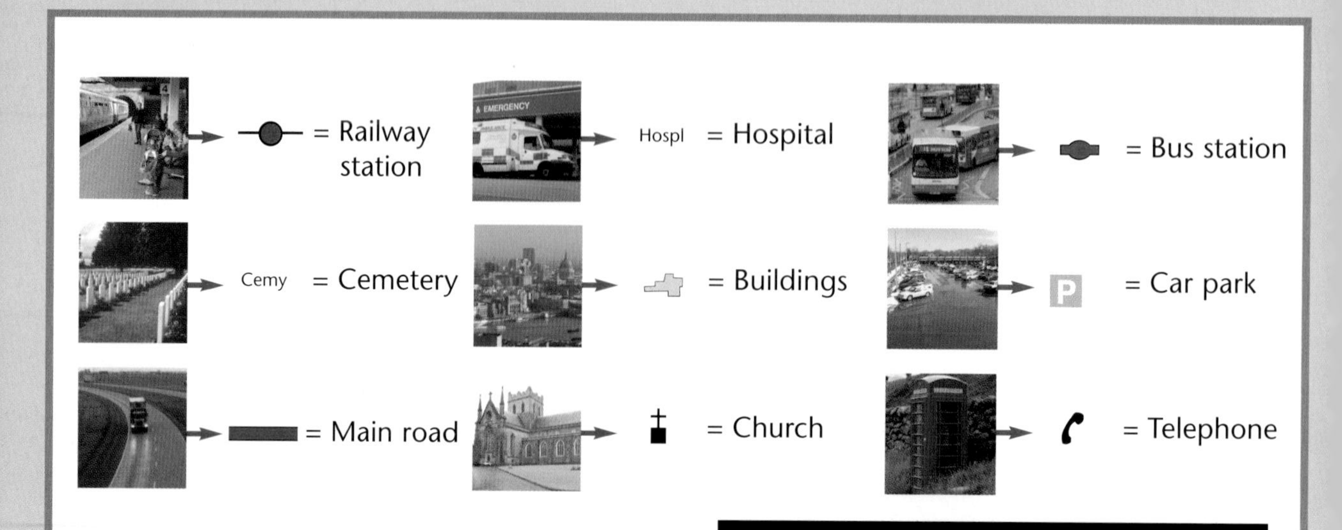

Here are some of the symbols that are used on many maps of settlements.

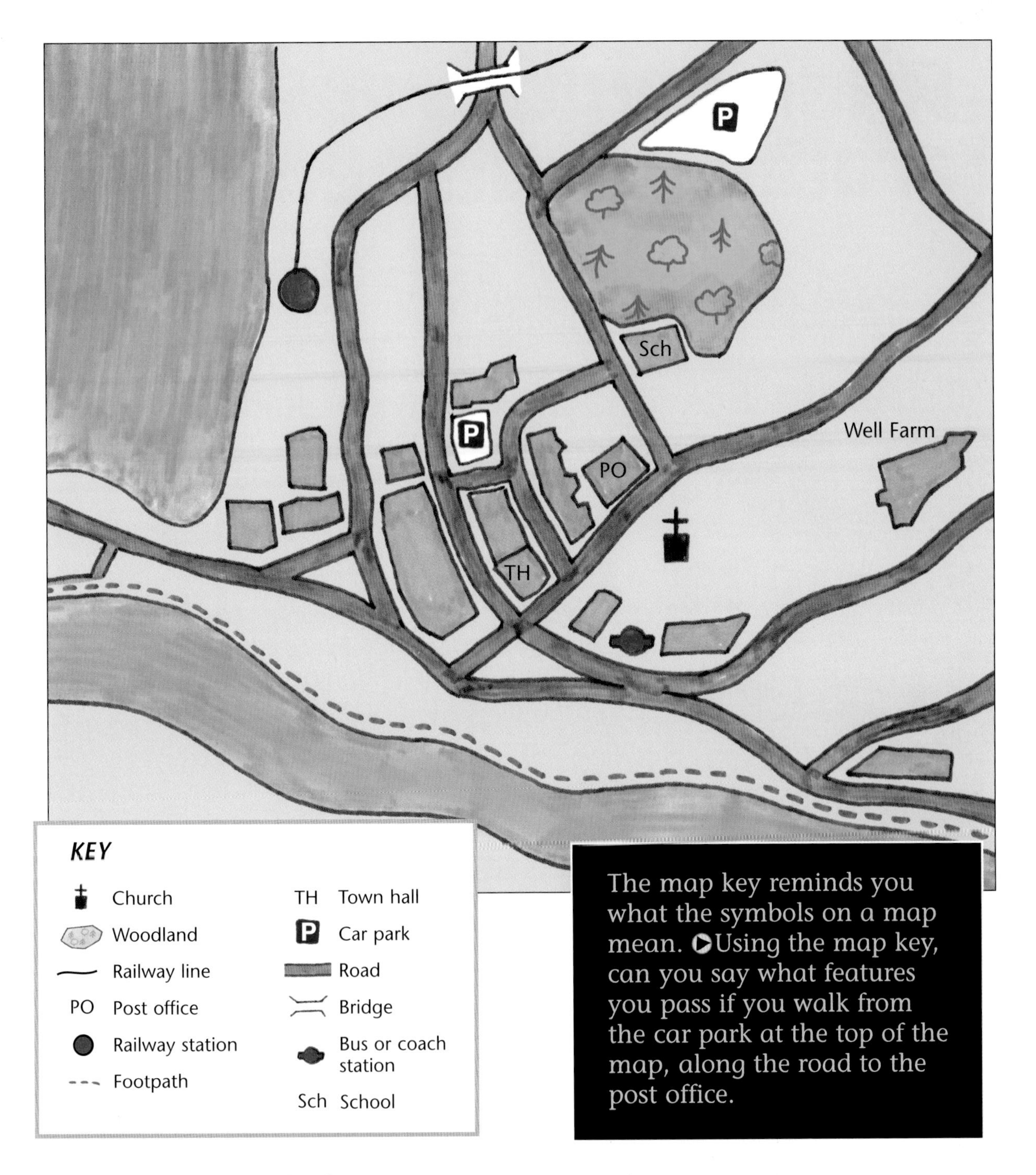

The map key reminds you what the symbols on a map mean. Using the map key, can you say what features you pass if you walk from the car park at the top of the map, along the road to the post office.

Transport symbols

When you look at a settlement map, one of the first things you notice are the transport networks. Railway lines, roads, canals, and cycle paths criss-cross villages, towns, and cities like spiders' webs. These different transport features are shown by different symbols. Other symbols that are linked to transport include bridges, car parks, and bus, coach, or train stations. Transport is a very important part of any settlement, because people need to get around and in and out of the places where they live.

Finding your way around settlements

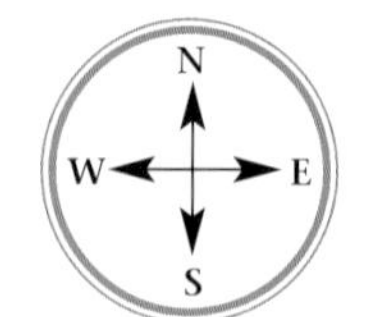

One way to locate places or features is by using compass directions – north, east, south, and west. Most maps are drawn with north at the top. This makes it easier to describe where something is by giving its direction from something else. For example, in the map below you could say that the wood is north of the post office and the church is south of the school.

Using grids

Many maps have lines that form a grid of squares. In the map below, the lines going up (vertically) have letters. The lines going across (horizontally) have numbers. A grid reference tells you which square a feature is in. For example, you can find the railway station in C,5 and the school in G,4. You can get more help with reading grid references on page 28.

▶ What is the name of the building in grid reference J,3? What is the grid reference for the church?

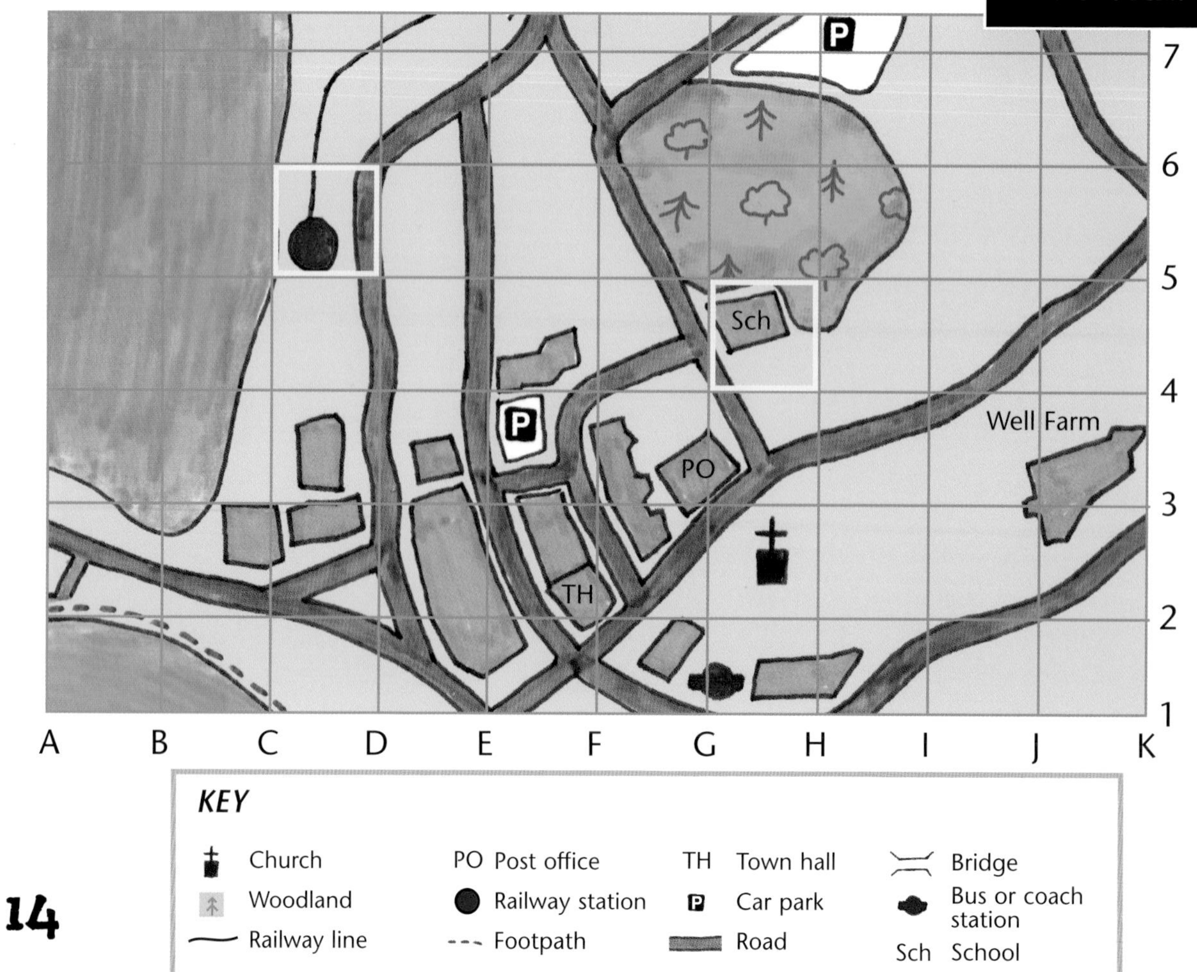

This is a map of the village of Lavenham in East Anglia. We have used yellow 'spotlights' to help you find things.

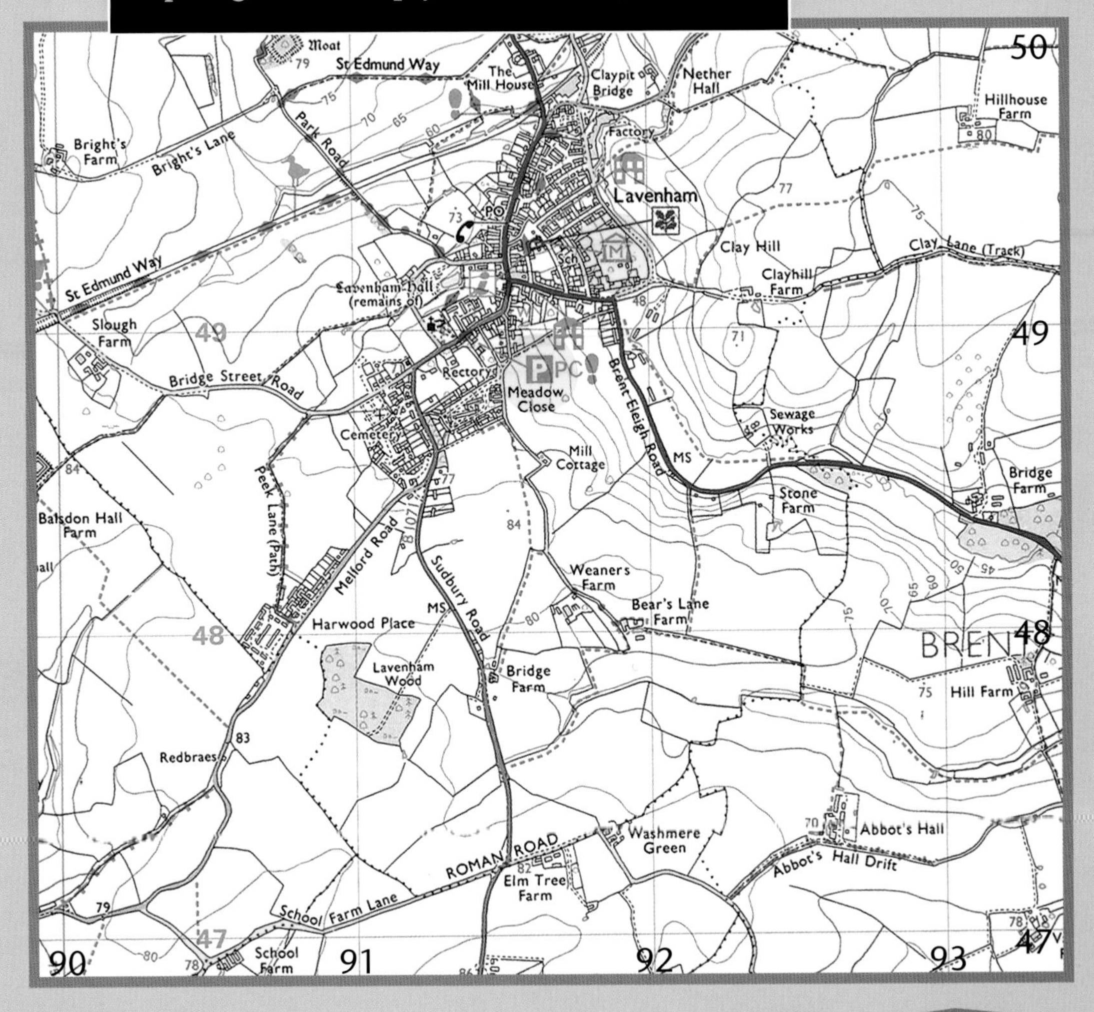

KEY

Car park

Museum

Nature reserve

Telephone

Walks

Grid games

The map on this page is a small part of an **Ordnance Survey** map. Ordnance Survey makes maps for the whole of the UK, and many people use them. Maps like this use numbers to label both the vertical and the horizontal grid lines. Grid references for these maps give the number of the vertical line first and then the number on the horizontal line. Remember this by the phrase: 'Along the corridor and up the stairs'.

In this map of the village of Lavenham, the car park is in square 91, 48.
▶ What is the grid reference for the museum?

Settlements high and low

Most settlements have been built on flat, low ground. This is because it is easier to build houses on flat ground. However, in the past many settlements were developed on hills or areas of high ground. This made them easier to defend from enemy attack. How can you tell from a map whether a settlement is on high, sloping, or flat ground?

Map contours

Many maps use contour lines to show the **gradient**, or steepness of a landscape. Contour lines join up land that is the same height above sea level. The diagram on this page shows you how contour lines work. The main thing to remember is that if the lines are drawn close together, it means the slope is steep. If the lines are farther apart, it means the slope is gentle. On some maps, the steepness of roads and paths is shown by gradient arrows, like this >>>. The more arrows, the steeper the climb!

The lower part of this diagram shows two mountains from the side. We have drawn contour lines up the mountains at every 50 metres. The contour lines wrap all the way round the mountains. The top part of the diagram shows you how the contour lines would look on a map.

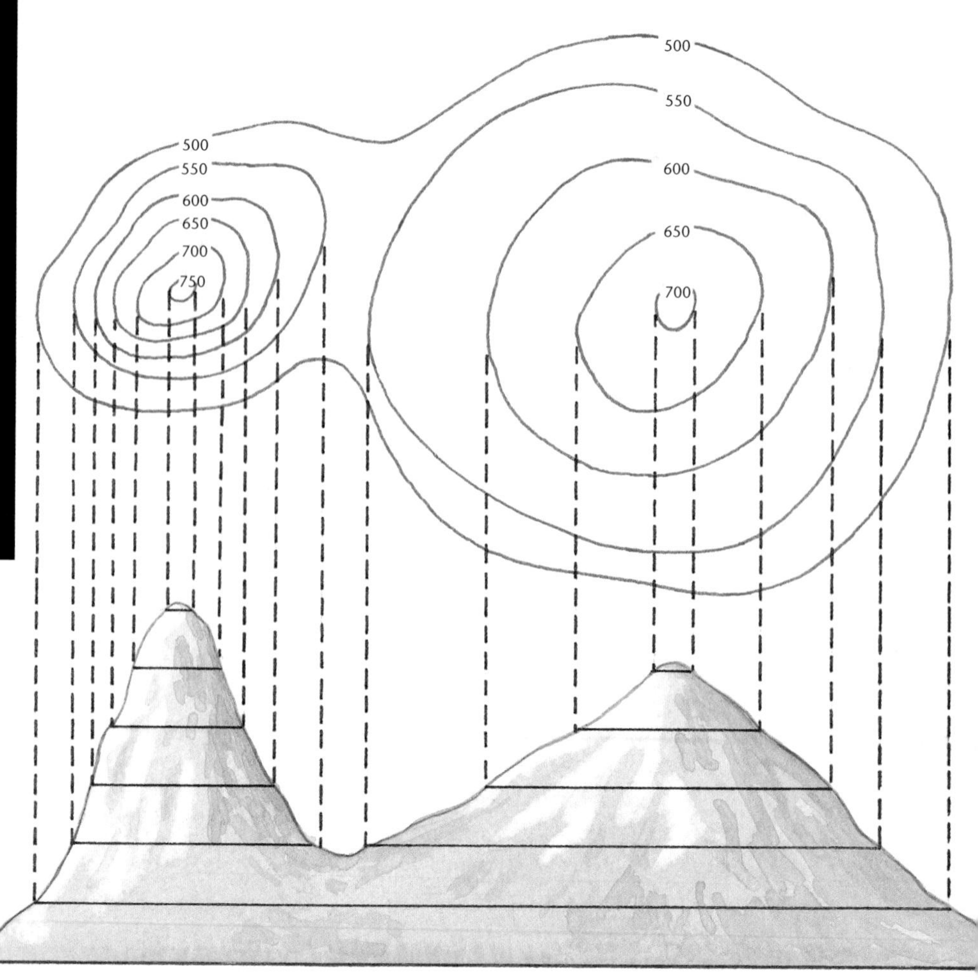

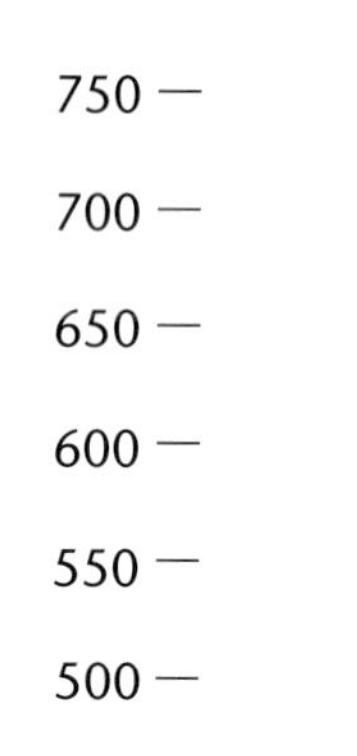

Many mountains and hills are formed from old **volcanoes.** Arthur's Seat is all that's left of a volcano that erupted around 350 million years ago!

Mapping Holyrood Park

The **Ordnance Survey** map on this page shows Holyrood Park, near the centre of Edinburgh in Scotland. We can see from the contour lines that Holyrood Park is on a steep hill. The highest point of the hill is called Arthur's Seat. The city of Edinburgh developed around the long road that runs between the Palace at the foot of this hill (in grid square 27, 73) and the castle to the west (in grid square 25, 73).

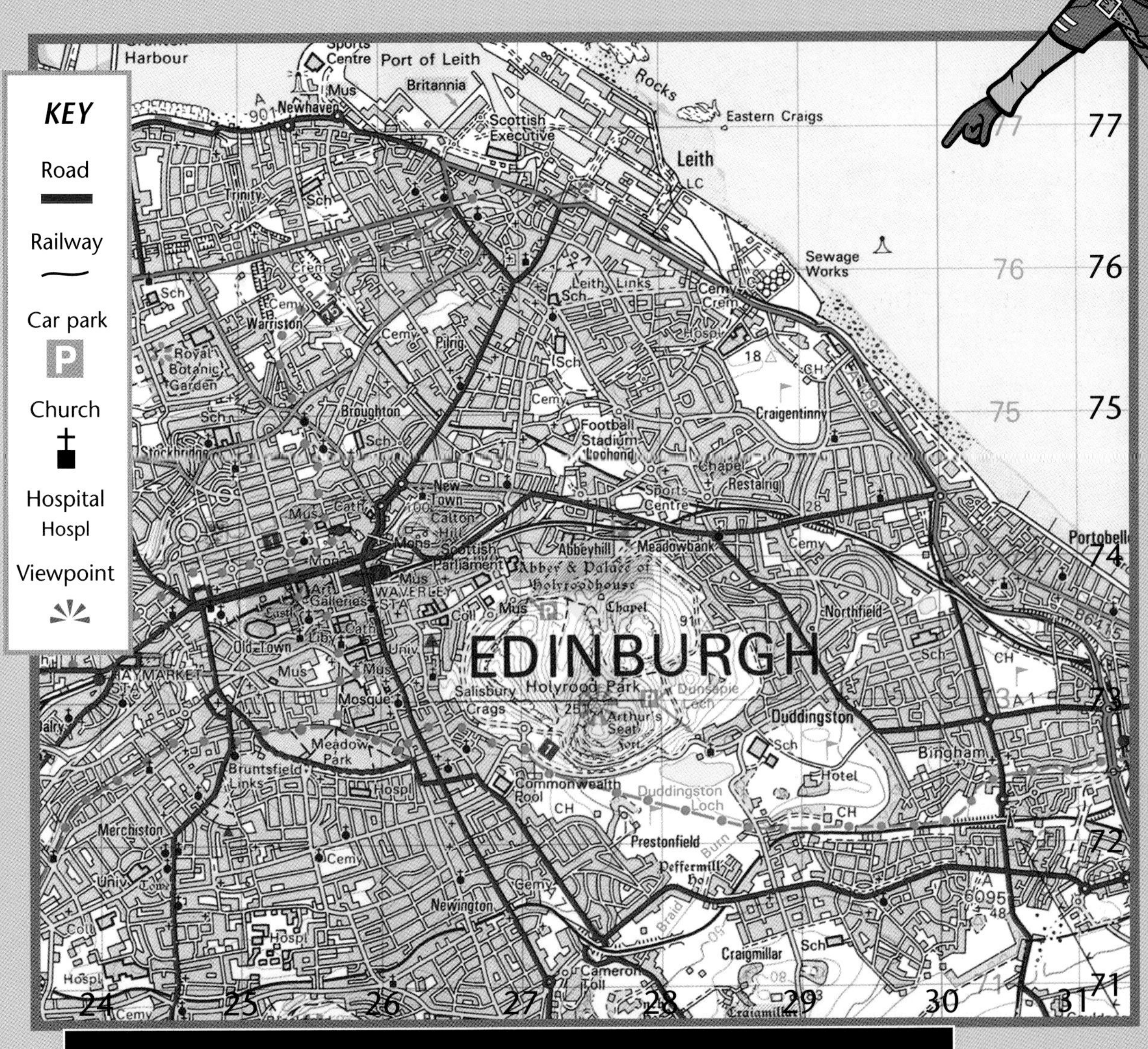

The contour lines on this map show us where the hill is. We can see that the hill is steep because the lines are close together. The yellow dotted line around the bottom of the hill is a path. From the top of the hill, you can get a good view of the city of Edinburgh. Can you see the viewpoint?

A day in Durham

Let's visit another city! Durham is a city in the north east of England in the county of Durham. People believe that its name came from two Anglo-Saxon (old English) words 'Dun Holm', meaning the 'hill island'.

Looking down on Durham

The 'hill island' is really a **peninsula**, an area of high ground surrounded on three sides by a river. The city of Durham developed around two important buildings on this hill – the cathedral and the castle. These were both built over 900 years ago. The site was chosen because it was high up and protected by the River Wear. This made it easier for people to defend their settlement from attackers. As Durham grew, houses and narrow winding streets were built on the land running downhill from the castle.

Can you see the cathedral in the middle of the peninsula? Now can you find it on the map opposite?

Mapping Durham

Compare the **aerial** photograph of Durham with this **Ordnance Survey** map of the same area. Can you find the cathedral and the castle on both the photograph and the map? (It might help you to know that the castle is now part of Durham University.) Other buildings on the map are shown as squares or rectangles. Some of these buildings have labels to say what they are. Some of the labels are shortened versions of a word. ▶ What do you think the labels *Coll*, *Sch*, and *Cemy* stand for?

▶ Using the map key, can you work out how many museums are shown on this map?

If you look carefully at the map, you can see the contour lines drawn very closely together around the inside edge of the peninsula. This shows us that it rises steeply above the surrounding land.

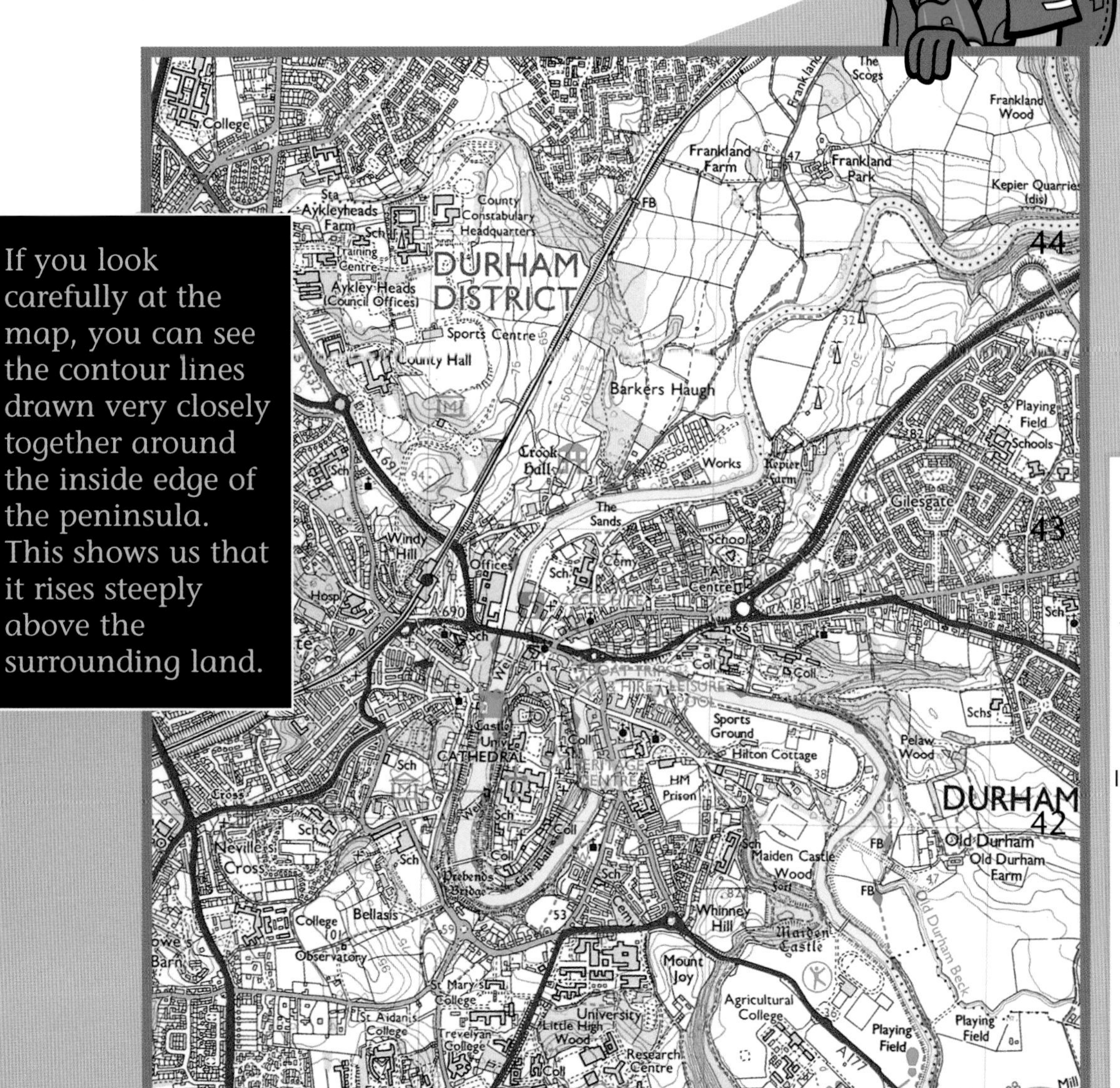

KEY

Church

Telephone

Railway

Information centre

Cathedral

Castle

Museum

Scale and settlements

It is pretty obvious that mapmakers have to draw things much smaller on maps than they are in real life! The scale of a map tells you exactly how much bigger something is in real life compared to its size on a map.

Sorting scales

Maps are drawn to different scales. A map scale of 1:25 000 means that 1 centimetre on the map is equal to 25,000 centimetres (or 250 metres) in real life. These big numbers are difficult to work with but you can also use the map grid to work out scale and distance. On **Ordnance Survey** maps, one side of a grid square always represents 1 kilometre on the ground. A map's scale is usually shown on a scale bar. You can see examples of scale bars on the bottom of the maps on these pages.

This is a 1:25 000 scale map of a village called Chillington in Devon. This kind of settlement is called a **linear settlement** because it follows the line of a road.

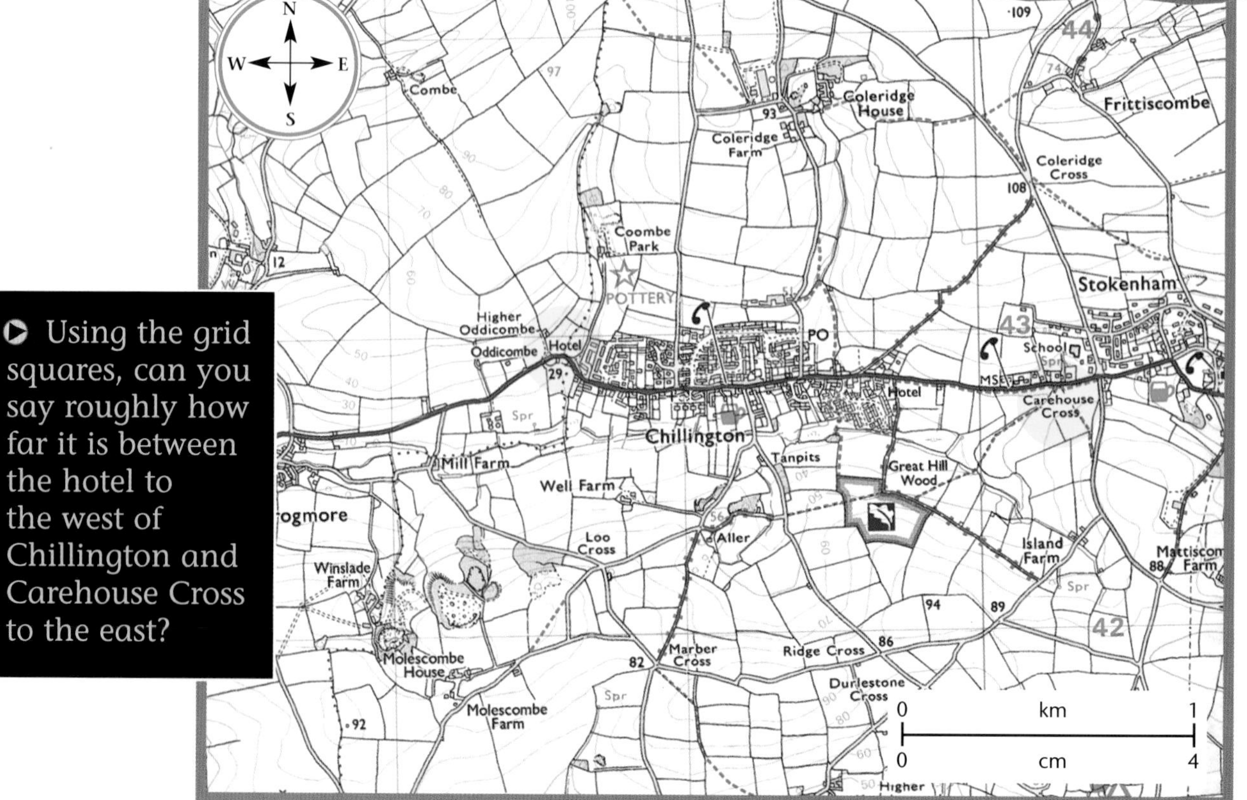

Using the grid squares, can you say roughly how far it is between the hotel to the west of Chillington and Carehouse Cross to the east?

Using different scales

Different scale maps are used for different purposes. A 1:25 000 map is a large-scale map. Large-scale maps show small areas of land in lots of detail. They are useful for walkers because they show features such as fields and footpaths. Maps with a scale of 1:50 000 or 1:100 000 are small-scale maps. They show a larger area, but without as much detail.

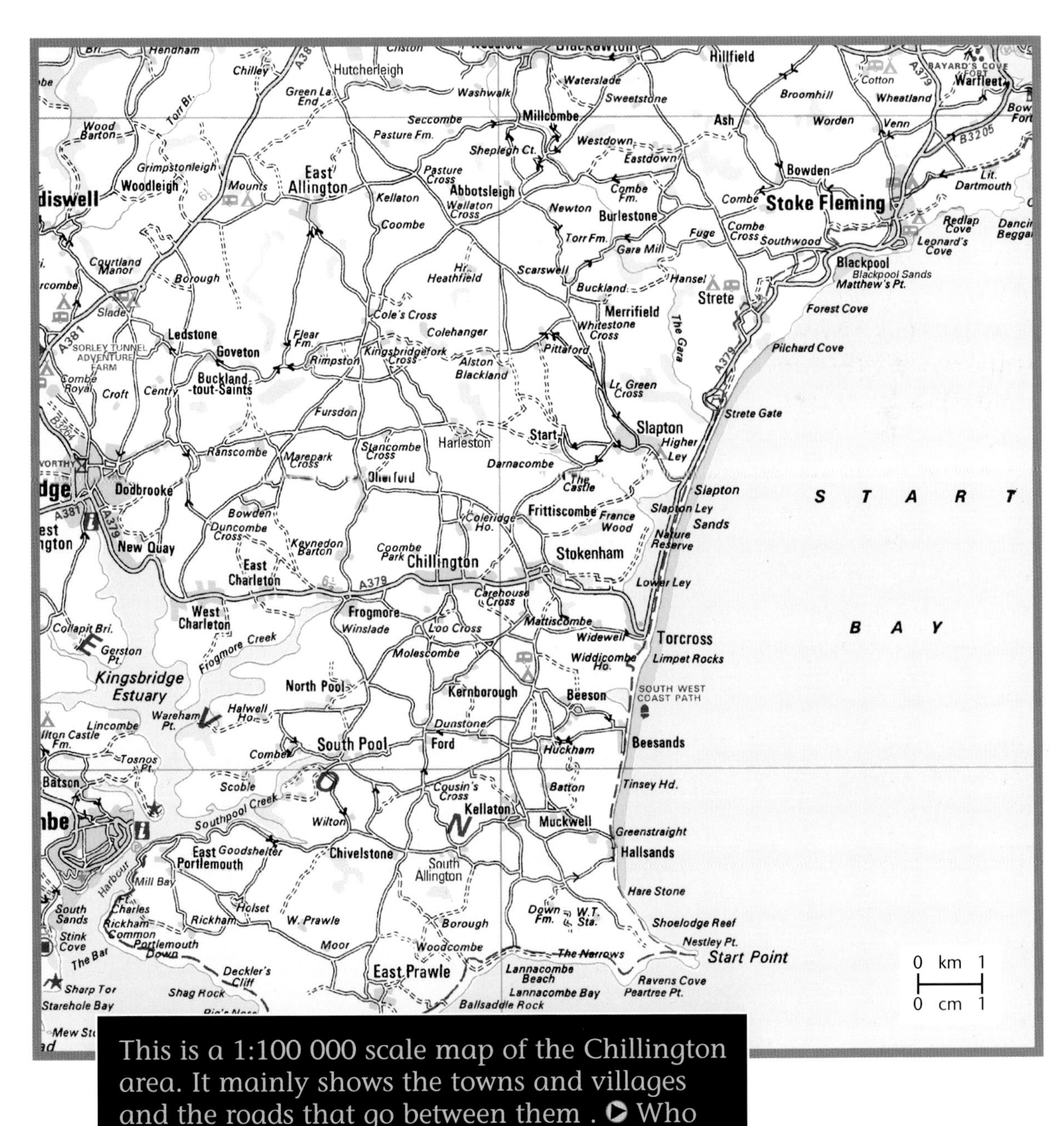

This is a 1:100 000 scale map of the Chillington area. It mainly shows the towns and villages and the roads that go between them . Who do you think would find this kind of map useful? Can you guess why?

Settlements of the UK

The map on this page shows the names of some of the larger settlements across the whole of the UK. Cities are the largest kinds of settlements in the UK. You can see on this map that there are more large cities in the mid and southern regions of the UK than in the north or west. The cities shown by red squares are capital cities. A capital city is where the government of a country is usually based. Can you find the area where you live on this map?

You can use the map scale to work out roughly how far one settlement is from another. ▶ Using the scale bar, can you work out how far it is to fly from Plymouth to London?

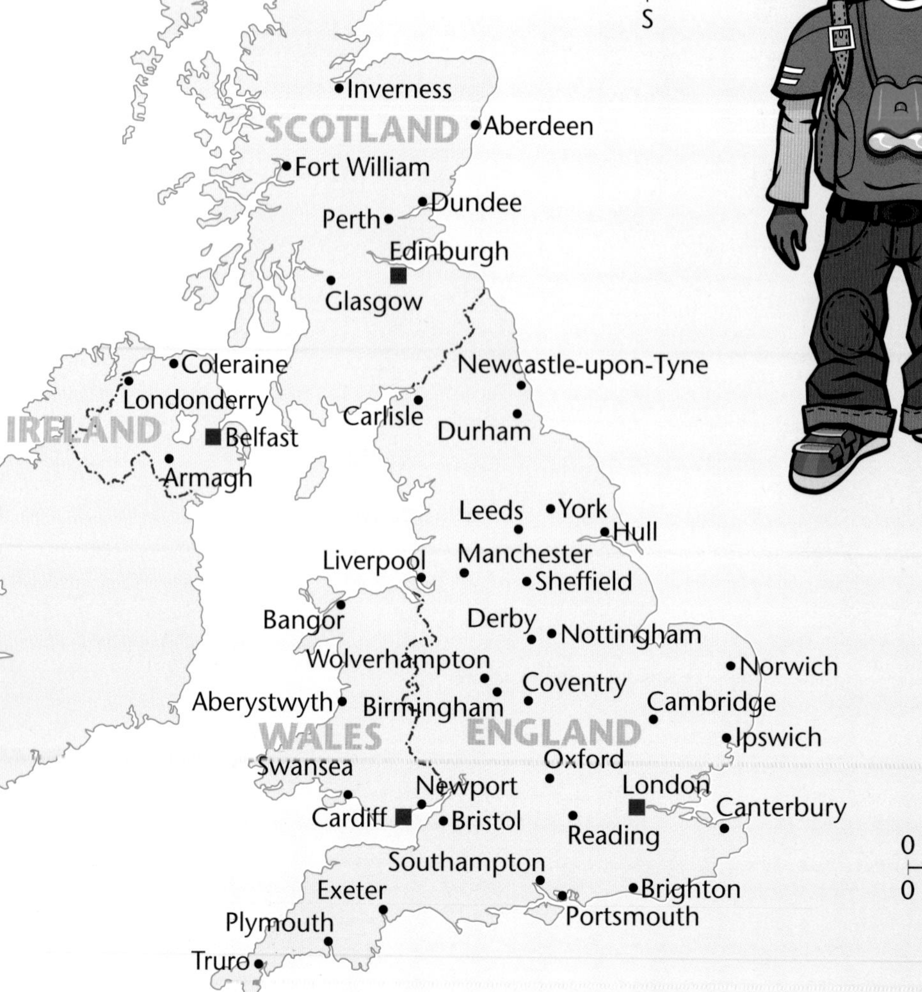

This is a **satellite** photograph of the British Isles at night. The lights show us just how many settlements there are across the country. The biggest cities give off the most light. The areas with less lights show that fewer people live there.

The trouble with traffic

Many of the lights you can see in the picture above are street lights. Roads in towns and cities can get very busy. Fumes from the vehicles' exhaust pipes cause noise and air **pollution**. Air pollution can give people breathing problems, such as asthma. Many cities have bypasses. These are roads that circle round a city to keep traffic out of busy town centres. The bypasses help **traffic congestion**, but some people object to them because they destroy areas of countryside and wildlife habitats. What do you think about this issue?

Going global

The Earth is a giant, rocky ball, and the most accurate way to map it is by using a globe. A globe is a ball-shaped model that shows the shape and sizes of the different parts of the Earth. However, globes are awkward to carry around so people also make flat maps of the world like the one shown on these pages.

This is a picture of a globe. You could try using a globe to find the UK, and some of the settlements shown on the flat world map.

The map here shows a range of settlements around the world. Most world maps, like this one, have labels for the **continents**,and oceans. This helps us to work out where places are in relation to each other.

Québec
NORTH AMERICA
New York
Los Angeles
ATLANTIC OCEAN
Mexico City

The cities with the biggest populations in the world are Tokyo in Japan, Mexico City in Mexico, and São Paulo in Brazil. Can you find them on the map?

SOUTH AMERICA
PACIFIC OCEAN
São Paulo
Rio de Janeir
Buenos Aires

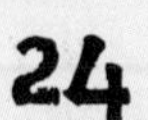

World settlements

Across the world, more and more people are moving away from countryside villages (**rural** areas) into large cities (**urban** areas) to find work. Also, big cities often have bigger and better **services**, such as hospitals. However, in many countries, big cities do not have enough space or jobs for extra people. This means that a great many people end up very poor, living in crowded **slums** on the edges of the cities.

ARCTIC OCEAN
London
Paris
EUROPE
Moscow
ASIA
PACIFIC OCEAN
Beijing
Seoul
Tokyo
Shanghai
Rabat
Tehran
Delhi
Karachi
Cairo
Dacca
Kolkata
Mumbai
Bangkok
Manila
AFRICA
Lagos
INDIAN OCEAN
Jakarta
AUSTRALIA
Sydney
SOUTH ATLANTIC OCEAN
SOUTHERN OCEAN
ANTARCTICA

This map shows some of the largest cities of the world. Can you see the cities marked by red stars? These cities have populations of over 10 million people.

Escape from Eden!

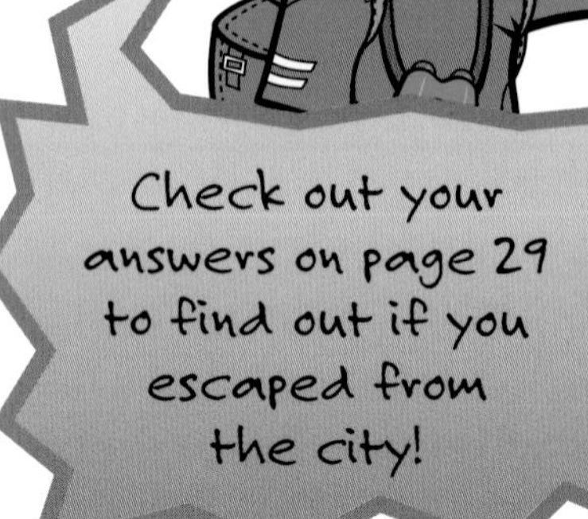

Are you ready to test your map skills? Imagine you are lost in the maze of streets in the city of Eden. The only way to get out is to follow the route given below to reach the Great Gate in the wall. Along the way, collect the first letters of the names of the features you meet. These will make a password that will open the gate!

Check out your answers on page 29 to find out if you escaped from the city!

1 Your journey begins at the Castle in the west of the city. Take the road leading east. Go over the crossroads. What do you come to?

2 You decide you would like to get off the road for a while. You head for grid reference 22, 43. What feature do you find here?

3 After a short rest, you head south for just over 1 kilometre. (Use the grid squares to measure this.) What is the name of the one-way street here?

4 Follow this street east and then north, to grid reference 23, 44. What is the name of the large grey building you come to?

5 Head north from here to the next feature. What does the key tell you this is?

6 From here, go west past two street turnings. What is the name of the feature you pass through next?

7 Follow the road past two shops, a blacksmith's, and a stable. What is the name of the big square that you come to?

8 Now follow the road heading east. You pass a large building with three roofs. What is it called?

9 Head south for 4 kilometres. What do you find in grid square 21, 41?

10 Now follow the road west that takes you past some houses, a shop, and an archery field. What is the name of the arch that you pass through here?

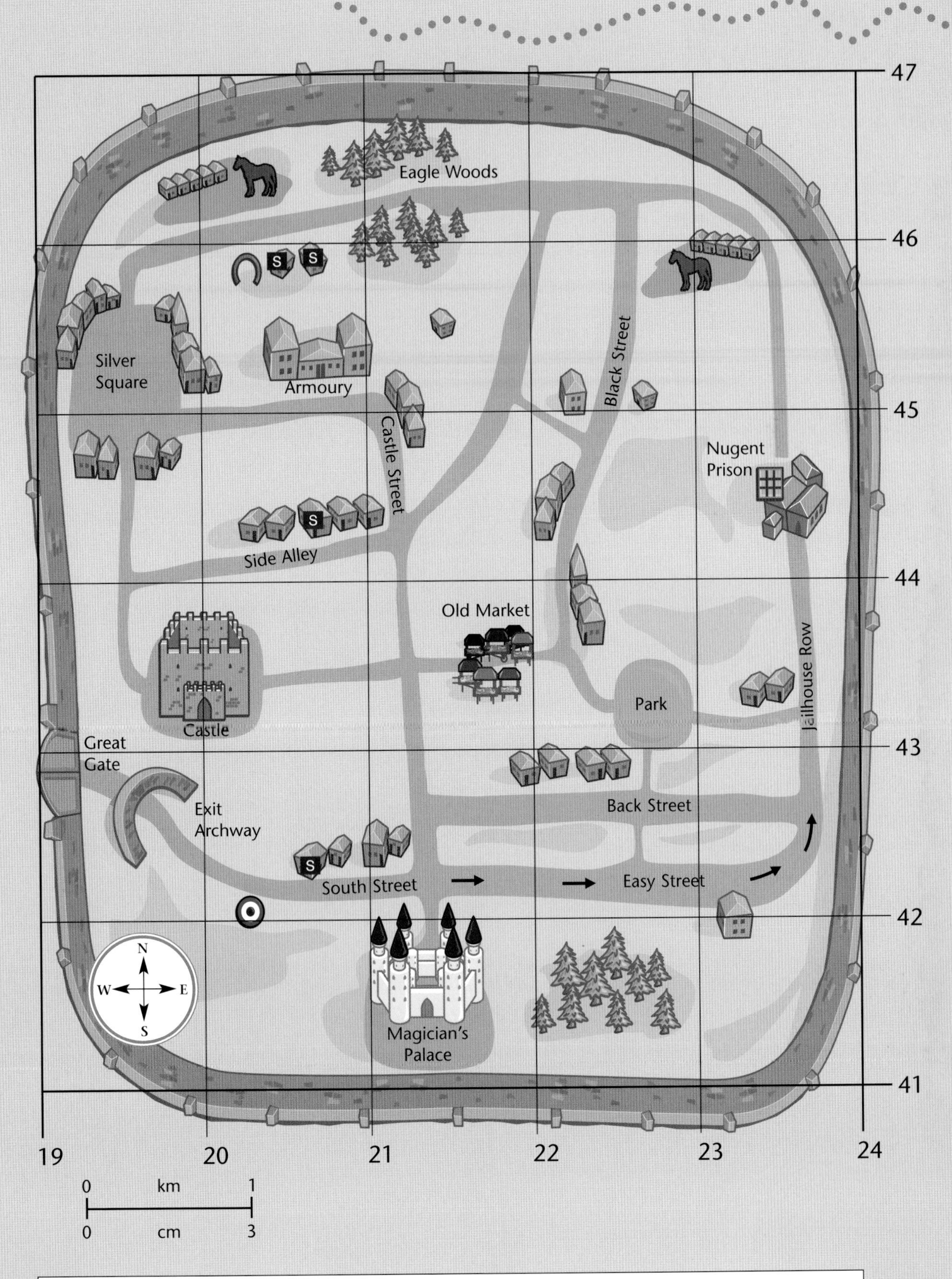
47
46
45
44
43
42
41
19
20
21
22
23
24
Eagle Woods
Silver Square
Armoury
Black Street
Castle Street
Nugent Prison
Side Alley
Old Market
Jailhouse Row
Castle
Park
Great Gate
Exit Archway
Back Street
South Street
Easy Street
Magician's Palace
N
W
E
S
0
km
1
0
cm
3

KEY
Blacksmiths
Shop
Stables
Archery field
Prison
Woods

Quick-stop map skills

What are symbols?

Map symbols are pictures, letters, shapes, lines, or patterns that represent different features, such as rivers and roads. Map keys show what the symbols stand for.

Key

Hospl = Hospital

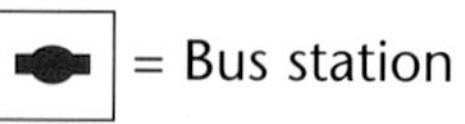

How can I measure distances?

On **Ordnance Survey** maps, each grid square represents 1 kilometre, so you can roughly judge distance by counting grid squares.

What are grid references?

Grid references are numbers that locate a particular square on a map. To give a grid reference, you give the number on the vertical line first and then the number on the horizontal line. ('Along the corridor and up the stairs.')

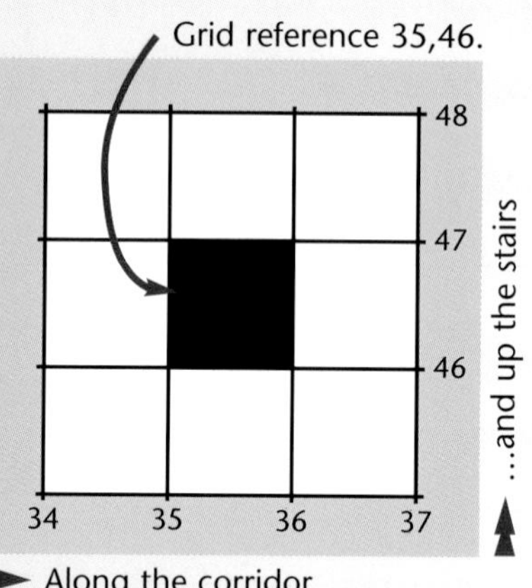

How does scale work?

A map scale tells you how much smaller a feature is on a map than it is in real life. Everything on a map is scaled down in size. On a 1:25 000 scale map, things are 25,000 times smaller than real life.

How are slopes shown on a map?

Maps show slopes by using shaded colours, contour lines, or gradient arrows. Gradient arrows look like this >>>. The more arrows, the steeper the slope is. Contour lines join up areas of land that are the same height above sea level. When contour lines are close together the slope is steep. When contour lines are spaced out, the land is flatter. Numbers next to, or on the lines tell us the exact height of the land in metres.

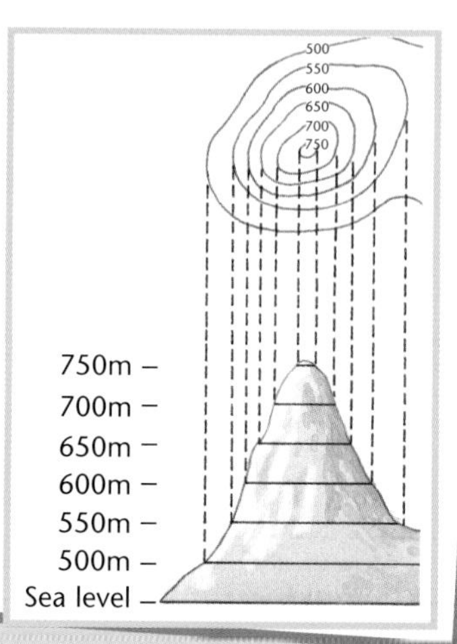

Map skills answers

Page 8: Clues from the photograph and map that suggest tourists might like to visit Tenby are the boats, beaches, and castle.

Page 11: On this section of the photomap of Cardiff you can see two sports clubs: Cardiff Athletic Club, on the left of the Millennium Stadium, and the Lawn Tennis Club in Cathays Park (on the bottom left side of the map).

Page 13: If you walk from the car park at the top of the map, along the road to the post office, you will pass a wood and a school.

Page 14: The name of the building in grid reference J,3 is Well Farm. The grid reference for the church is G,2.

Page 15: The grid reference for the museum is 91, 49.

Page 19: *Coll*=College, *Sch*=School and *Cemy*=Cemetery. We can see three museums in this section of the map of Durham. These are in the following grid squares: 27, 43; 26, 42; and 26, 40.

Page 20: The distance from the hotel (to the west of Chillington) to the crossroads called Carehouse Cross (to the east of Chillington) is approximately 1.75 kilometres.

Page 21: Drivers would find small-scale maps useful. Because these maps show a larger area of land, they show more towns and villages – and more of the major roads that lead to them.

Page 22: It is approximately 320 kilometres (200 miles) to fly in a straight line from Plymouth to London. It would be further to drive between these two cities because the roads would not be straight.

Page 26: The password you need to escape the city of Eden is OPEN SESAME! 1 – O (Old Market); 2 – P (Park); 3 – E (Easy Street); 4 – N (Nugent Prison); 5 – S (Stables); 6 – E (Eagle Woods); 7 – S (Silver Square); 8 – A (Armoury); 9 – M (Magician's Palace);10 – E (Exit Archway).

Glossary

aerial overhead, from the sky

continents largest land masses in the world. Each continent is divided into different countries.

dispersed settlement settlement where buildings are spread apart from each other

gradient the steepness of a slope

harbour safe place at the coast where boats can come in from the sea

industry type of work. Industries include fishing, mining, tourism and making cars in factories.

linear settlement settlement that has developed along a line, for example along a road or river

mouth end of a river, where it flows into the sea

nucleated settlement settlement where the buildings are all built around a centre

Ordnance Survey map-making organization that makes maps that cover the whole of the UK

peninsula piece of land almost surrounded by river or sea water

permanent features things that are always in the same place. A car park is a permanent feature; a car parked there is not.

pollution when water, air or soil are made dirty or poisonous by people's waste

rural countryside area

satellite scientific object in Space that can send out TV signals or take photographs

service job or business that supplies people in settlements with something that they need, such as electricity or water

site where a settlement has been built

slums overcrowded part of a settlement where the houses are dirty and unhealthy

traffic congestion when there are too many cars on a road

tourism industry that provides holidaymakers with things that they need, such as ice-creams or rooms in a hotel

urban area of a town or city

volcano mountain or hill made by hot liquid rock, called lava, that erupted from inside the Earth

Find out more

Books

Wild Habitats: Towns and Cities, Louise and Richard Spilsbury (Heinemann Library, 2004)

Philip's Junior School Atlas (4th edn), (Heinemann, Rigby, Ginn, 2003)

Websites

You can play games, get homework help and learn more about using Ordnance Survey maps on the Ordnance Survey mapzone site:
www.ordnancesurvey.co.uk/mapzone

The global eye website has lots of information about global cities. This includes problems such as air pollution. There is also a satellite image of the world at night so you can see the lights of settlements across the globe!
www.globaleye.org.uk/secondary

By entering place names or postcodes, you can see aerial photographs and maps of particular places and at different scales:
www.multimap.co.uk

Check out the geography section on the Heinemann Explore website to find out even more about maps!
www.heinemannexplore.co.uk

Index